The Theory of Spin

The Theory of Spin

in planets and atoms

By

Keith Dixon-Roche

with a little help from

Isaac Newton

The Theory of Spin

Keith Dixon-Roche © 2017

All concepts and formulas in this book

are the sole property of Keith Dixon-Roche

and protected by copyright.

Their use, publication, broadcasting,

distribution, copying or any form of recording

without Keith Dixon-Roche's written consent

shall be a breach of international copyright law

and subject to legal action.

The Theory of Spin

in planets and atoms

Published by CalQlata
info@CalQlata.com

First published December 2019
Final publication April 2026

This book is sold subject to condition
that it shall not by way of trade or otherwise,
be lent, re-sold, hired out or otherwise circulated
without the publisher's prior consent and in such
circumstances it shall not be circulated in any form of
binding or cover other than that in which it is published

Contents

Preface		7
1 Introduction		9
	1.1 Goodricke & Algol	11
	1.2 Chicken & Egg	12
2 Energy		13
	2.1 Electrical	14
	2.2 Magnetic	17
	2.3 Potential	20
	2.4 Kinetic	21
3 Polar Moment of Inertia		23
	3.1 Formulas	24
4 Celestial Spin		25
	4.1 Component Energy E_0	27
	4.1.1 Formulas	28
	4.2 Component Energy E_1	29
	4.2.1 Formulas	30
	4.3 Total Energy E_2	31
	4.3.1 Formulas	32
	4.4 Component Energy E_3	33
	4.4.1 Formulas	35
	4.5 Verification	36
	4.5.1 Formulas	37
	4.6 Core-Mantle Rotation	40
	4.6.1 Formulas	45
	4.7 No moon	46
	4.8 Magnetic Fields	47
	4.8.1 Formulas	48
	4.9 Magnetic Reversal	49

5	Atomic Spin		51
	5.1 Formulas		52
6	Spin in Our Galaxy & Solar System		55
	6.1 Celestial Spin		56
		6.1.1 Hades	58
		6.1.2 Sun	59
		6.1.3 Mercury	60
		6.1.4 Venus	61
		6.1.5 Earth	62
		6.1.6 Mars	64
		6.1.7 Asteroid-Belt (Ceres)	65
		6.1.8 Jupiter	66
		6.1.9 Saturn	67
		6.1.10 Uranus	68
		6.1.11 Neptune	69
		6.1.12 Pluto	70
		6.1.13 Moon	71
	6.2 Atomic Spin		72
Appendices			75
	A1 References		77
	A2 Glossary		79
	A3 Symbols		81
	A4 Orbital Motion (formulas)		83
	A5 Useful Formulas		85

The Theory of Spin

Preface

Over three-hundred years ago; together with the help of Johannes Kepler & Galileo Galilei, Isaac Newton derived *'Principia'; the* definitive mathematical theory of universal orbits.

But it's more important than that:
It is the origin for the story of our universe.
However, it was missing one critical feature; *spin.*
The purpose of this book is to fill this gap and complete Newton's story.

Together, orbits and spin provide us with the only true answer to *'life, the universe and everything'*; energy.

We can't blame Newton for omitting spin from his work, as it would have been impossible with the information and facilities available to him. It is a pity, however, that nobody since has taken the time to resolve it, because if they had, we would not now be deluged with such nonsense as *'Relativity'* and *'Quantum Theory'*.

These two theories (orbits and spin) not only tell us everything we need to know about our universe (from the atom to the Big-Bang), but also where we can find limitless clean energy. For example; less than one centimetre of the earth's dry crust contains enough clean energy for the rest of this universal period, and it is free and safe, none of the fuel will need storing or recycling!

The distribution of inefficient and dirty energy sources has been the focus for politicians and industrialists to make their fortunes and impose their control over you and me. So, I decided that it was about time Newton's heroic work was completed, making energy available universally; and here it is:

Spin Theory mathematics defines the energy driving angular motion in all matter, and it is induced by Newton's orbits.

Because it is now possible to define the Milky-Way's force-centre, I have given it the name 'Hades' for easier reference.

Keith Dixon-Roche 2026

The Theory of Spin

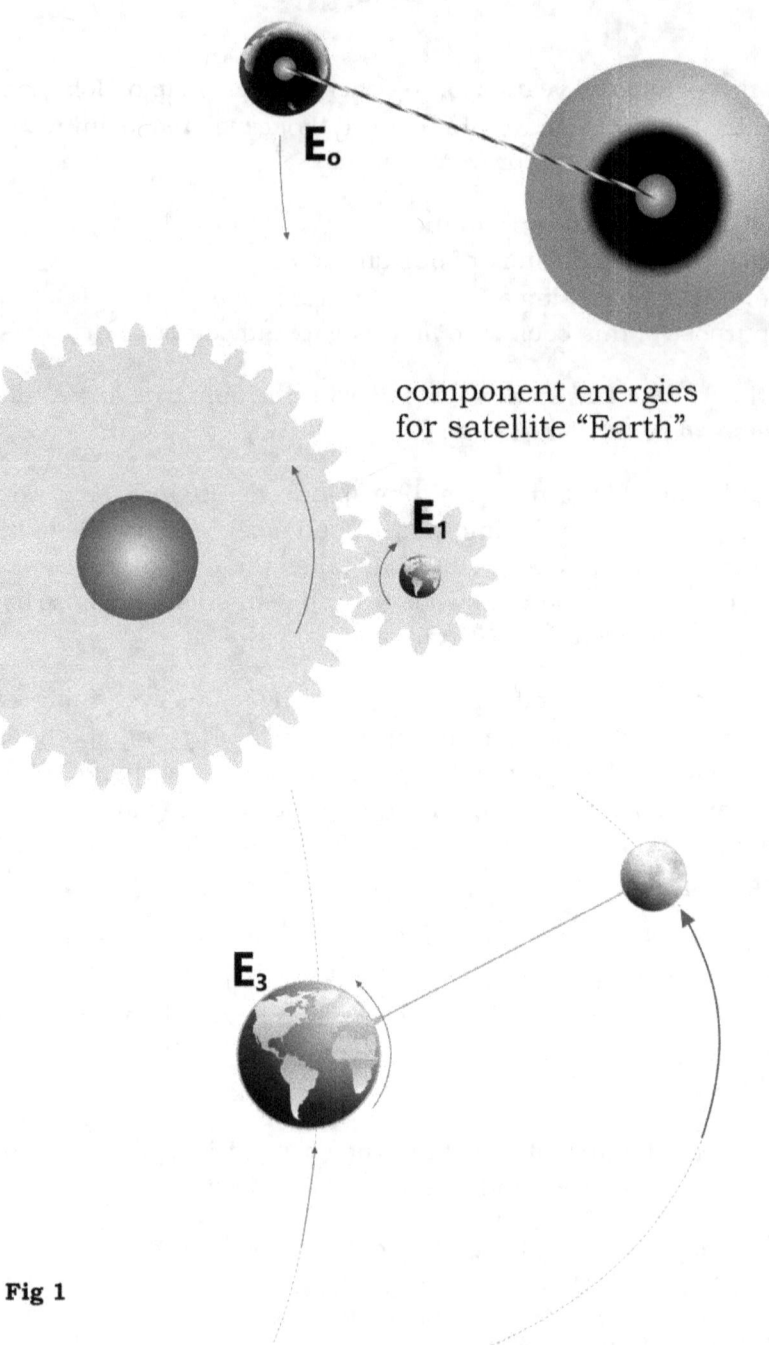

component energies for satellite "Earth"

Fig 1

1 Introduction

Spin is the angular motion of a body about an axis passing through its centre of mass. Its energy is generated by orbits. This is a fundamental law of nature.

There are three mechanisms generating component energies that drive spin in all satellites.
These are described in Fig 1, in which the satellite is a planet, its force-centre is a star and its sub-satellite is a moon:

E_0 is the spin energy induced in the satellite's core by the potential energy between it and its force-centre.

E_1 is the spin energy induced in the satellite' mantle by the angular spin energy in its force-centre.

E_3 is the spin energy induced in the satellite's mantle by the kinetic energy of is sub-satellite(s).

Total spin energy in a satellite is the sum of these component energies.

These differential spin rates, in a satellite's core and its mantle are responsible for creating its internal frictional heat.

The heat energy induced by spin builds up over time. Once it matches the satellite's radiated heat energy, the satellite's temperature will stabilise.

If the power (energy per unit time) generated by a satellite's sub-satellite population (mass) is sufficient, the satellite's crust will melt. This is what creates gas planets. On the other hand, bright stars are gas planets, with sufficient satellite population to achieve the temperature needed to create neutrons and thereby release fissionable energy.

The reason spin is so important to our lives is that it generates the energy that enables life to exist. If there were no spin, stars and planets would be cold and dark, and there would be no neutrons, so no energy would be stored for the next '*Big-Bang*'; i.e. no universal period.

The Theory of Spin

Planetary Spin

The kinetic energy in planets and stars is induced during a '*Big-Bang*' event. The difference between planetary kinetic (+ve) and potential (-ve) energy is responsible for generating angular motion in all celestial bodies. Whilst space remains empty, the orbits generated by celestial potential and kinetic energies will remain constant between '*Big-Bangs*' (perpetual motion).

Atomic spin

All satellites that orbit in circular paths around their force-centres, such as electrons and man-made satellites, *must* provide their own kinetic energy. The potential energy between a force-centre and a satellite in such an orbit is always twice the satellite's kinetic energy.

1.1 Goodricke & Algol

In 1784, John Goodricke discovered the [supposed] binary nature of the star originally named Algol. As one of the binary stars passed in front of the other, their combined brightness dimmed, revealing two important facts:

1) One of the stars was bright (hot) and the other was dark (cold)

2) Only the bright star was a force-centre for an orbital system

The heat generated by the *bright* star is due to its dedicated satellite population. The other darker star (or large planet) is actually in orbit around the bright-star but has few satellites of its own. As such, the dark partner can generate little internal heat.

Once a star (force-centre) has acquired its satellites, it can trap a twin but it will not share its satellites; the twin will orbit the star.

This discovery reinforces the fact that bright stars generate their internal heat from their satellite population.

1.2　Chicken & Egg

Which came first; spin or orbit?

What you see in most films and documentaries is that the sun starts spinning and the planets follow it around. This is of course 'back-to-front'.

In order to generate spin, you need an appropriate energy. Spin theory teaches us that if a sun, planet or moon sat alone in space it would not spin.

Spin was first induced in our sun by the rotational energy in its force-centre (Hades; ≈1E-15 radians per second), and it would have continued to spin at this rate had it not acquired satellites (planets) of its own. However, our sun actually rotates much faster than this; >2.865E-06 radians per second.

If angular kinetic energy in a force-centre induces orbital kinetic energy in its satellites, this transfer of energy would slow down the force-centre's rotation, which is obviously not the case. I.e. kinetic energy in planets must induce rotational kinetic energy in their star.

Therefore, our planets must have been orbiting long before our sun achieved rotation anywhere near its current rate.

The same argument applies to a spiral galaxy. Our sun got its initial spin from the spin energy in Hades, but Hades is in a linear orbit. Therefore, all of Hades's spin must have been induced by its orbiting satellites.

So, orbits came first!

2 Energy

Energy is possessed by matter in the form of electrical and magnetic charge. It is generated in electrical or magnetic fields, and transferred by electro-magnetic radiation. It exists in kinetic or potential form.

With regard to celestial spin; magnetic energy ties bodies together (potential energy or gravity), electrical energy generates their heat (kinetic energy or friction) and the relative rotary motion of the electrical and magnetic charges generates a satellite's magnetic field. This is borne out by the fact that the greater a satellite's internal size and heat, the greater its magnetic field.

2.1 Electrical

The application of potential energy (potential difference) sufficient to pull electrons from one atomic shell and transfer them to another along a conductor is what we currently refer to as DC (direct current) electricity.

Artificially induced electricity is generated by moving an electrical charge within a magnetic charge (e.g. in motors and generators) and is what we currently refer to as AC (alternating current – electrical field).

Electrical energy is shared between particles and travels from negative towards positive

The Theory of Spin

2.1.1 Charge

The static electrical charge (e) can be negative (electrons) or positive (protons). It is ever-present in all Quanta and is all-pervasive. It radiates in all directions and its attraction/repulsion is felt by all other Quanta throughout the universe. It retains magnitude irrespective of distance.

It is a myth that the force induced by an electrical charge fades with the square of the distance from its source as this would conflict with the conservation of energy. It is simply distributed over a larger spherical area.

Electrical charge is 2.269E+39 times magnetic charge, but because it is **shared** between Quanta:
the electrical attraction/repulsion between each of a million Quantum neighbours is one-millionth that of the electrical attraction between two Quanta. This is the reason we see negligible electrical attraction between celestial bodies (planets, stars, moons, etc.).

The electrical charge capacity (e or e_n) is defined by particle *mass* (m_e or m_p). The minimum magnitude (e) is always present and constant in all Quanta irrespective of circumstances.

A proton's additional mass provides it with the capacity to hold onto an additional electrical charge (e') commensurate with its electron's kinetic energy while it is part of a proton-electron pair. This additional charge is collected from its electron partner via the opposing static electrical charges and used by the proton to generate (and emit) electro-magnetic energy and electrical field energy.

The operational electrical charge (e') is the energy that repels adjacent atoms (gas).

When a proton loses its electron, its electrical charge will fall to that of the electron; the elementary charge unit (e).

According to Coulomb, electrical potential energy is calculated thus:
$PE = k.q_1.q_2/R$

But because electricity is shared, q_1 & q_2 in this calculation must be equal to the lesser of the two values; e.g. in a proton-electron pair:

$PE = k.e^2/R$

2.1.2 Field

Electrical field energy is generated by proton-electron pairs and varies with proton electrical charge (e'). It acts as the carrier for the electro-magnetic energy radiated by a proton-electron pair.

Field electricity is artificially induced. It can be generated by bringing together two atoms (or collections of atoms) of strong opposite polarity (DC), or forced into a conductor by rotating a magnetic field (AC) such as passing an electrical conductor through a rotating magnetic charge (e.g. in motors and generators).

Field electricity is not addressed in detail in this book because it plays no part in spin theory.

2.2 **Magnetic**

Magnetic energy is accrued between particles and travels from positive towards negative.

Artificially induced magnetism is polar and it is generated either by encircling a bar magnet within an electrical field (e.g. transformers) or moving a magnetic charge within an electrical charge (e.g. planetary field).

Magnetism is what we currently understand as mass and gravity.

2.2.1 Charge (Mass & Gravity)

Magnetic charge is non-polar magnetic energy present and constant in all Quanta. It is what we understand today as mass, and the field lines it radiates in all directions are what we understand as gravity.

It is a myth that the force induced by magnetic charge fades with the square of the distance from its source as this would conflict with the first law of thermodynamics. It is simply distributed over a larger spherical area.

Magnetic energy is 4.407E-40 times electrical energy, but because it is **accrued** between Quanta:
the magnetic strength of 100 Quanta is 100 times stronger than in one Quantum. This is why planets and stars - that comprise enormous numbers of Quanta - have such a strong magnetic attraction, but is so much weaker in, say, a cup.

The relative strength of the magnetic charge is defined by Quantum *mass* (m_e or m_p). It is always present and constant irrespective of circumstances. The magnetic charge in a proton is always m_p/m_e times greater than that of an electron.

If its shell-1 proton-electron pairs are (or can be) aligned within a body according to to its elemental lattice structure, they will generate strong magnetic polarisation. Whilst such alignment in iron (with a BCC lattice structure) must be induced artificially, e.g. with a lodestone, it occurs naturally in metals with an HCP structure such as Cobalt, Dysprosium, Gadolinium, Neodymium and Terbium. This is the magnetism that exists in bar-magnets.

Electro-magnetic radiation is deflected by magnetic charge.

According to Newton (and Gilbert), magnetic potential energy is calculated thus:
$PE = G.m_1.m_2/R$

And because magnetism is accrued, m_1 & m_2 in this calculation may be any value; e.g. in a proton-electron pair:

$PE = G.m_e.m_p/R$

2.2.2 Field

Magnetic field energy is generated by proton-electron pairs and constant. It is invariable because the proton magnetic charge remains constant irrespective of electron energy. This is the energy that attracts adjacent atoms (viscous matter) and the magnetism that is generated by the earth.

Field magnetism is artificially induced. It is generated by encircling magnetic matter within an electrical charge (as in transformers). It generates polar field lines that flow from one end to the other.

This form of magnetism is much stronger than the magnetic charge, but it is highly concentrated, unidirectional and has a very limited field of influence.

Polar magnetic fields are highly selective in the materials they attract because of their dependency on nucleic alignment (lattice structure).

Other than that generated by the earth (refer to chapter 4.8), field magnetism is not addressed in detail in this book because it plays no part in spin theory.

2.3 Potential

Potential (static) energy exists between all Quanta, irrespective of separation distance in the form of what we currently understand as gravity but is actually due to the combined magnetic charge in all the Quanta in any and all bodies.

Potential energy is the attraction or repulsion of either electrical or magnetic energy.

Negative potential energy (PE: gravity) holds particles together and positive potential energy (CE: centrifugal) keeps them apart.

In a balanced system (e.g. orbits), both PE and CE must be equal.

PE = m.a.R

Newton's potential energy is calculated thus: PE = $G.m_1.m_2/R$

In circular orbits:

PE = 2.KE = 2 . ½.m.v^2 = m.v^2 (Henri Poincaré)

2.4 Kinetic

Kinetic (dynamic) energy, which exists in all moving particles, is always positive and transferred via electrical, magnetic, electro-magnetic or impact [potential] energy.

Electro-magnetic energy is absorbed by orbiting electrons and converted into kinetic energy.

Kinetic energy in elliptical orbits is generated and maintained by the varying potential energy between the satellite and its force-centre.

Kinetic energy in a satellite following a circular orbit (such as in an atom) is not induced into the satellite by its force-centre such as in elliptical orbits; it must be provided by the satellite itself.

$KE = \tfrac{1}{2}.m.v^2$ (Henri Poincaré)

The Theory of Spin

3 Polar Moment of Inertia

Polar moment of inertia is a body's resistance to being spun on its axis.

If you know a body's angular velocity (ω) you can calculate its radial modifier (Δ). Alternatively, if you know its radial modifier, you can calculate its angular velocity.

When used in conjunction with Core Pressure, 'Δ' can help us to determine a body's internal structure.

Below is listed the polar moment of inertia and radial modifier (Δ) for each of the planets in our solar system:

Celestial Body	Δ	J (kg.m^2)
The Sun	0.318281959	3.900014E+46
Mercury	0.812862196	5.193084E+35
Venus	0.68123191	3.309127E+37
Earth	0.334460097	1.080860E+37
Mars	0.002317087	1.583269E+31
Jupiter	**0.02278067**	**1.925856E+39**
Saturn	**0.014060011**	**1.523954E+38**
Uranus	**0.024937619**	**1.389062E+37**
Neptune	**0.065237927**	**1.056968E+38**
Pluto	8.64241985	5.484999E+35
Table 3.1		

The above 'Δ' values for the **gas planets** are, of course incorrect; they are based upon outside radii and angular rotation rate of the planet's cloud cover (atmosphere). An alternative calculation has been carried out for these planets based upon more representative body density (5392 kg/m^3), the results for which are provided below.

Celestial Body	Δ	J (kg.m^2)
Jupiter	0.036358991	1.925855E+39
Saturn	0.027939584	1.523892E+38
Uranus	0.040523774	1.399249E+37
Neptune	0.096393201	1.042762E+38
Table 3.1.1		

As expected, the polar moment of inertia is identical to the cloud-covered planet (Table 3.1)

The above values are also based upon the angular velocity of the planet's cloud cover. This issue is addressed in Chapters 6.1.8 to 6.1.11

The Theory of Spin

3.1 Formulas

Polar moment of inertia of a homogeneous mass (Fig 2) is calculated thus:

$J = ⅖.m.r^2$

Where:
'm' is the body's mass
'r' is its outside radius

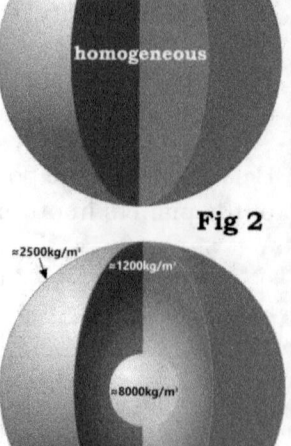

Fig 2

However, this formula only works if the body comprises a mass of constant density, which is not the case for celestial bodies, such as; planets, stars, moons, etc. (Fig 2). Gravity tends to ensure that the denser matter migrates towards their cores.

This problem can be solved by using a radial modifier (Δ) thus:

$J = ⅖.m.(\Delta.r)^2$

The greater the Δ value the lower the density variation;

$\Delta < 1$ means the core density is greater than its surface density (normal situation)

$\Delta = 1$ means that the entire planet is homogeneous

$\Delta > 1$ indicates that the body is being pulled into a local orbit

In the case of planets with a dense atmosphere (such as the gas planets), we do not know their dimensions or their spin rate, so we cannot accurately calculate their density (Table 3.1).

We can, however, estimate their radial modifier based upon the orbital energy (E_o) and planetary mass assuming an average *planetary* density of the innermost planets:

Surface radius: $r = \sqrt[3]{[\ 3.m\ /\ 4.π.ρ\ ]}$ ($ρ = 5392$ kg/m³)

Radial modifier: $\Delta = \sqrt{[\ 5.E_o/m\ ]}\ /\ (r.\omega_o)$ (Table 3.1.1)

4 Celestial Spin

Planetary spin is an indisputable fact of life. That is why we have days and years here on earth. But what causes it?

All astro-physicists today claim that either there is no force spinning the planets or that it is impossible to calculate owing to the chaotic nature of the solar system. Both these views are incorrect. Planetary spin can be easily predicted with the same degree of accuracy as Newton's laws of orbital motion from the potential and kinetic energies calculated using them.

Magnetic [potential] energy (gravity) and satellite kinetic energy are together responsible for generating it, but only between a satellite, its force-centre and its sub-satellites.

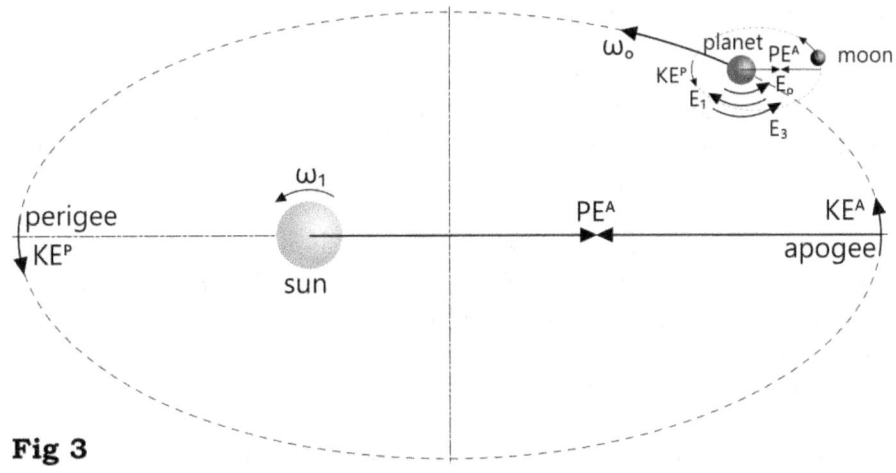

Fig 3

Fig 3: Potential energy between a force-centre and that of its satellite will naturally cause the satellite to spin at the same angular velocity and in the same rotational direction (prograde) as its orbit around the force-centre (ω_0). However, spin energy induced in the satellite by the force-centre's own rotational kinetic energy will cause the satellite to rotate in the opposite (retrograde) direction ($-\omega_1$).

If a planet's moon is orbiting in the same rotational direction as the planet's orbit e.g. prograde, the moon's kinetic and potential energies will cause the planet to rotate also in a prograde direction (ω_1).

The Theory of Spin

Spin theory is not as difficult or complex as everybody appears to believe.

For example: Venus's opposite rotational direction cannot, apparently, be explained, but the reason for it is quite simple once you understand spin theory.

E_3, which is generated by a planet's moon(s), is always considerably greater than E_1 and E_0 (Fig 3) and therefore defines a planet's rotational direction. As '$E_3 = 0$' in a planet with no moon, its own orbital energies (E_1 & E_0) define its spin direction. Being a (relatively) large planet, E_1 is the dominant factor in Venus's spin direction.

The same argument applies to Mercury except in its case its smaller size means that E_0 is dominant so it spins in the same direction as the other planets in our solar system.

A significant difference will occur between the angular velocity of a planet's core and its mantle only if it has a substantial moon.

The only unknowns you need in order to calculate planetary spin are the polar moment of inertia of the planet and that of its sun. When calculating the spin in a planet, you need either its angular velocity (ω) or its radial modifier (Δ). If you have one, you can calculate the other.

The competing rotational energies (E_1, E_3 & E_0) along with their different distribution (@ the core or throughout) are responsible for the conflicting angular velocities in a planet's core and mantle and thereby generating its internal heat. This rotational velocity difference in the earth's core and mantle is also responsible for generating the earth's magnetic field.

This calculation method is as accurate as Newton's own laws of motion and is essentially an extension of them. Therefore, not only is planetary spin predictable, it is both simple and accurate.

Using this calculation method, it has been possible to determine what would happen to the earth's spin if it lost its moon (refer to Chapter 4.7).

Gas planets exist as such because they have been able to attract sufficient satellite *mass* to melt their [surface] crusts through internal friction. Unlike a bright star, however, they do not generate the heat required to create neutrons. It is also probable that the heat lost to a gas planet's heavy atmosphere will form a surface skin.

The Theory of Spin

4.1 Component Energy E_0

E_0 is the natural rotational energy developed in a satellite, assuming it presents the same face to its force-centre. It may be represented by the rotation that would be expected in a ball swung about your head, at the end of a length of string. The direction of spin induced in a satellite by 'E_0' will be the same as that of the satellite's orbit. The spin-period will be identical to the satellite's orbital period.

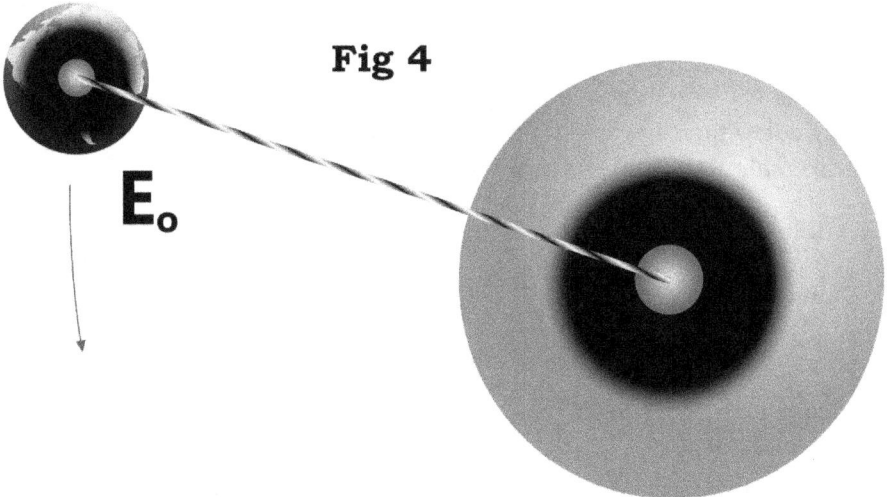

Fig 4

E_0

Magnetic charge (potential) energy, always acts between the centres of masses.

We know this to be true because:

a) Newton told us so, and;

b) It has never been disputed in 300 years.

Moreover, our solar system's only solitary satellite with no sub-satellites of its own (the earth's moon) does exactly this; our moon always presents the same face to the earth.

The Theory of Spin

4.1.1 Formulas

The angular rotation rate (ω_o) for a satellite due to its orbital period is calculated thus:

$$\omega_o = 2\pi / t_o$$

Where; 't_o' is the satellite's orbital period in seconds

The energy required to generate 'ω_o' is calculated thus:

$$E_0 = \tfrac{1}{2}.J.\omega_o|\omega_o|$$

Where: 'J' is the satellite's polar moment of inertia. This value was originally calculated iteratively, by hand, but has been verified by an alternative calculation method (refer to Chapter 4.5)

The results for our solar system are listed below:

Celestial Body	E_0 (J)	ω_o (°/s)
The Sun	1.461276E+16	8.65661E-16
Mercury	1.774469E+23	8.26678E-07
Venus	1.733075E+24	3.23644E-07
Earth	2.142284E+23	1.99099E-07
Mars	8.871154E+16	1.05859E-07
Jupiter	2.712883E+23	1.67849E-08
Saturn	3.481068E+21	6.75905E-09
Uranus	3.900860E+19	2.36992E-09
Neptune	7.714895E+19	1.20823E-09
Pluto	1.768504E+17	8.03026E-10
Table 4.1.1		

4.2 Component Energy E_1

E_1 is the rotational energy induced in a satellite by the rotational energy in its force-centre. This rotation may be represented by a pair of gears and behaves in the same way; i.e. the direction of spin induced in the satellite by 'E_1' will be opposite to that of the satellite's orbit.

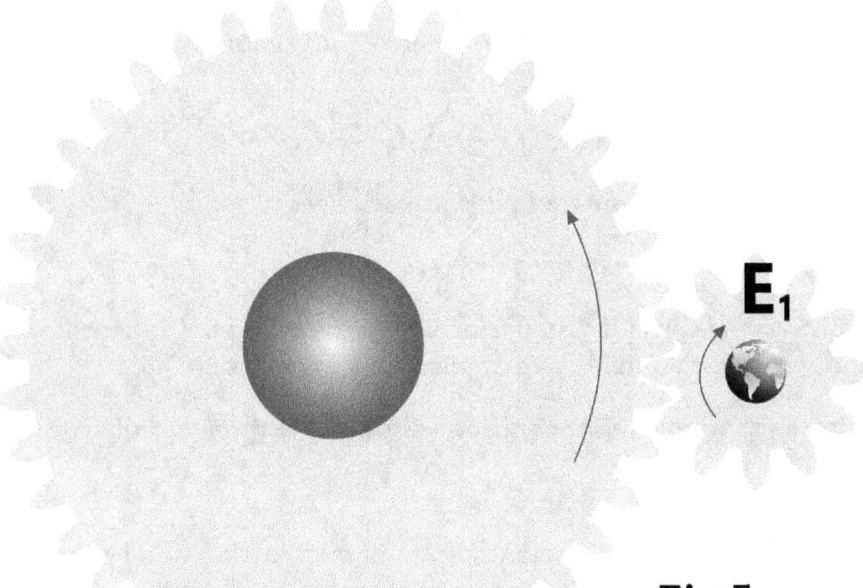

Fig 5

Because angular energy in a satellite (E_1) is significantly less than kinetic energy induced by sub-satellites (E_3), we would expect the relative angular motion of satellites with sub-satellites to reflect this.
And this is what we find:

	Mercury	Venus	Earth	Jupiter
ω_1/ω_2	**1.202**	**0.412**	3.34E-03	2.914E-03
	Saturn	Uranus	Neptune	Pluto
ω_1/ω_2	2.121E-03	6.258E-04	5.849E-05	1.328E-05

Mercury and Venus have no sub-satellites. ω_1/ω_2 for our sun is 3.79E-10

Mars is a special case. Because it is hollow, its spin ratio should be much higher than expected, which it is (0.625).

The Theory of Spin

4.2.1 Formulas

The rotational energy (E_1) induced in a satellite by the rotational rate of its force-centre is calculated thus:

$$E_1 = \delta KE \cdot (r/R)^2$$

Where;

'δKE' is the difference between the satellite's maximum and minimum orbital kinetic energies: $\delta KE = KE^P - KE^A$

'KE^P' is the satellite's kinetic energy at its perigee

'KE^A' is the satellite's kinetic energy at its apogee

'r' is the outside radius of the satellite

'R' is the average orbital distance between the centres of the satellite and its force-centre (half the length of the orbital major axis)

The satellite's rotational rate attributed to this energy is calculated thus:

$$\omega_1 = \sqrt{[\,2 \cdot E_1 / J\,]}$$

Where; 'J' is the satellite's polar moment of inertia. This value was originally calculated iteratively, by hand, but has been verified by an alternative calculation method (refer to Chapter 4.5)

The results for our solar system are listed below:

Celestial Body	E_1 (J)	ω_1 (c/s)
The Sun	2.300141E+16	1.086073E-15
Mercury	5.765629E+23	1.490135E-06
Venus	2.515333E+23	1.232980E-07
Earth	3.200840E+23	2.433673E-07
Mars	1.556136E+22	4.433650E-05
Jupiter	2.528419E+26	5.124219E-07
Saturn	9.191740E+24	3.473221E-07
Uranus	2.807922E+22	6.358382E-08
Neptune	2.093597E+21	6.294054E-09
Pluto	6.271063E+15	1.512158E-10

Table 4.2.1

4.3 Total Energy E_2

E_2 is the final (total) rotational energy in the satellite; i.e. the sum of all the other rotational energies.

We know this to be correct, because:

a) If E_0, E_1 & E_3 are correct, their sum should provide accurate results for all the satellites and force-centres in our solar system; which it does.

b) 170 orbital spin calculations have been performed in the making of this book, all of which reflect exactly what we see.

c) a calculator now exists, based upon this theory, that can predict the spin-rate of any and every celestial body in our solar system.

4.3.1 Formulas

The total rotational energy driving the rotation of a planet is the sum of the component energies, and is calculated thus:

$E_2 = E_1 - E_0 - E_3$

Where:

Component energy 'E_0' is calculated in Chapter 4.1

Component energy 'E_1' is calculated in Chapter 4.2

Component energy 'E_3' is calculated in Chapter 4.4

The angular rotation rate of the body is calculated thus:

$\omega_2 = \sqrt{[\,2.E_2/J\,]}$

Where: 'J' is the satellite's polar moment of inertia. This value was originally calculated iteratively, by hand, but has been verified by an alternative calculation method (refer to Chapter 4.5)

The results for our solar system are listed below:

Celestial Body	E_2 (J)	ω_2 (°/s)
The Sun	1.600977E+35	2.865329E-06
Mercury	3.991161E+23	1.239801E-06
Venus	-1.481541E+24	-2.992369E-07
Earth	2.877018E+28	7.296281E-05
Mars	3.977416E+22	7.088236E-05
Jupiter	2.977768E+31	1.758525E-04
Saturn	2.044087E+30	1.637867E-04
Uranus	-7.118296E+28	-1.012377E-04
Neptune	6.202913E+29	1.083382E-04
Pluto	-3.555149E+25	-1.138559E-05
Table 4.3.1		

4.4 Component Energy E_3

E_3 is the rotational energy induced in a satellite by its own sub-satellite(s). The direction of spin induced in the satellite by 'E_3' will be the same as the orbital direction and kinetic energies in these sub-satellites.

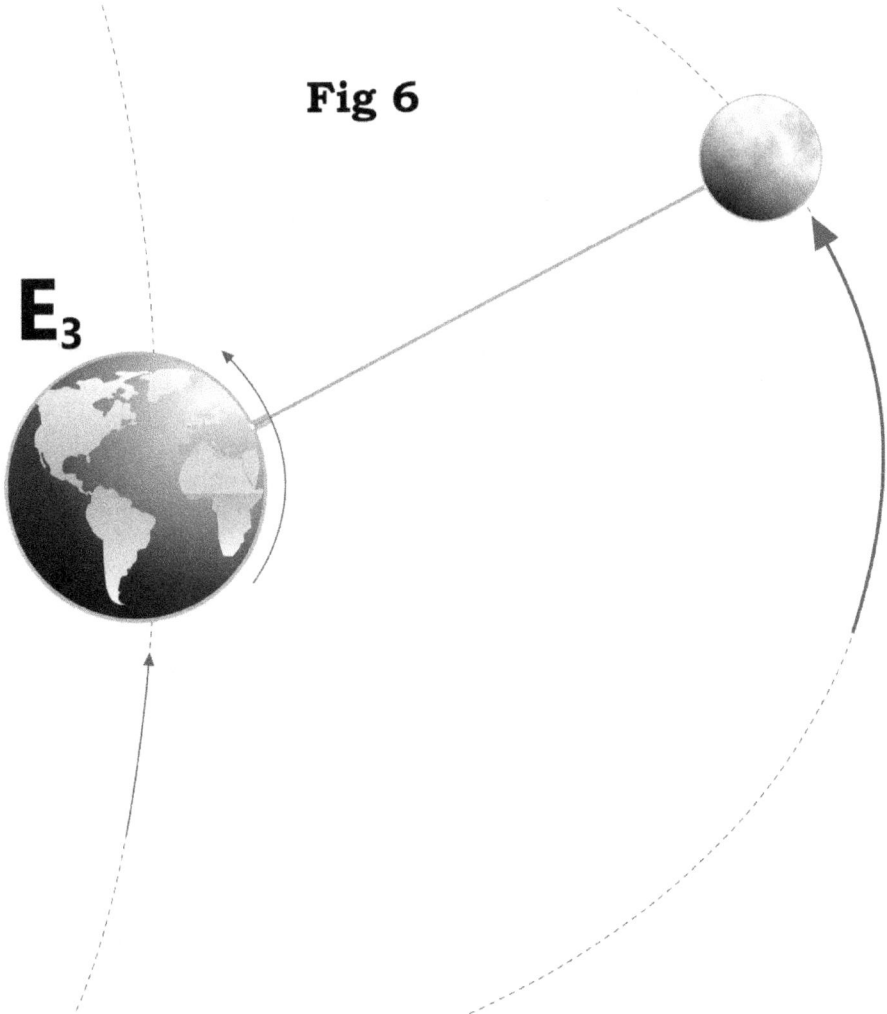

Fig 6

We know this because Newton told us so; i.e. that there is an attractive force between all masses, that he called *gravity*. Multiply this force by a separation distance and you get potential energy. If you move one such mass relative to another, its kinetic energy will induce a torque in the stationary mass via their potential energy (E_3: spin).

The Theory of Spin

Moreover, because of the hold Hades has over our sun's core (refer to Table 6.1.2), the fact that the sun's *surface* is spinning at the rate calculated using its planetary kinetic energy, E_3 *must* apply either at its surface or throughout its mass. In either case, the same calculation method applies.

Throughout our solar system, where a satellite hosts sub-satellite(s), E_3 is always dominant, and this is reflected in the remnant velocity ratios[#]:

[#] *'remnant velocity ratio' refers to the velocity ratio for force-centre induction*

	Mercury	Venus	Earth	Jupiter
$1 - \omega_3/\omega_2$	1	1	1.848E-06	4.241E-06
	Saturn	Uranus	Neptune	Pluto
$1 - \omega_3/\omega_2$	2.248E-06	1.955E-07	1.648E-09	2.399E-09

Mercury and Venus have no sub-satellites.

The potential energy induced in our sun by Hades, is so high that its core rotation rate is virtually the same as that of its orbit (refer to Table 6.1.2). The angular rotation we see *must* therefore, be driven almost entirely by the kinetic energy induced by its satellites. So, for our sun; $1 - \omega_3/\omega_2 \approx 0$ Which is exactly as expected when compared with Mercury and Venus (1).

Mars is a special case. Because it is hollow, we would expect its remnant velocity ratio to be significantly higher, which it is (0.22).

The Theory of Spin

4.4.1 Formulas

The rotational energy (E_3) induced in a satellite by its own sub-satellite's (moons) is calculated thus:

$$E_3 = \text{sign}(\cos(\theta)) \cdot (\Sigma KE^P + \Sigma PE^A)$$

Where;
'KE^P' is a sub-satellite's kinetic energy at its perigee
'PE^A' is the potential energy between a satellite and its sub-satellite at its apogee
'sign($\cos(\theta)$)' is either '1' or '-1' and reflects the tilt angle of the sub-satellite's orbital plane relative to the satellite's orbital plane

Notes: KE^P & PE^A are for the sub-satellite(s)
PE is always negative
'KE' is positive if the sub-satellite is orbiting in the same direction as the satellite and negative if it is orbiting in the opposite direction.

The satellite's rotational rate attributed to this energy is calculated thus:

$$\omega_3 = \sqrt{[\ 2 \cdot E_3 / J\ ]}$$

Where; 'J' is the satellite's polar moment of inertia. This value was originally calculated iteratively, by hand, but has been verified by an alternative calculation method (refer to Chapter 4.5)

The results for our solar system are listed below:

Celestial Body	E_3 (J)	ω_3 (°/s)
The Sun	-1.600977E+35	-2.865329E-06
Mercury	0	0
Venus	0	0
Earth	-2.877008E+28	-7.296268E-05
Mars	-2.421289E+22	-5.530457E-05
Jupiter	-2.977743E+31	-1.758518E-04
Saturn	-2.044077E+30	-1.637864E-04
Uranus	7.118298E+28	1.012377E-04
Neptune	-6.202913E+29	-1.083382E-04
Pluto	3.555149E+25	1.138559E-05 |//
Table 4.4.1		

4.5 Verification

The calculations shown in Chapters 4.1 to 4.4 have been verified by an alternative calculation procedure that reveals exactly the same results.

To calculate the radial modifier of a satellite, you need to know its angular rotation rate.

To calculate the angular rotation rate of a satellite, you need to know its radial modifier.

If you have one, you can calculate the other.

We know the angular rotation rates of the planets in our solar system, so we can calculate the radial modifier for each planet.

The Theory of Spin

4.5.1 Formulas

$E_1 = \delta KE \cdot (r/R)^2$
Where:
δKE is the satellite, 'r' is the satellite's body radius and 'R' is the mean satellite's orbital radius

$E_3 = \text{sign}(\cos(\theta)) \cdot (\Sigma KE^P + \Sigma PE^A)$
Where: PE is always negative and KE^P & PE^A are for the sub-satellite(s)
'sign(cos(θ))' is either '1' or '-1' and reflects the tilt angle of the sub-satellite's orbital plane relative to the satellite's orbital plane

$\omega_0 = 2\pi / t_o$

$\omega_2 = 2\pi / t_s$

Where;
t_o is the satellite's orbital period
t_s is the satellite's spin (angular rotation) period

$J = 2 \cdot (E_1 - E_3) / (\omega_o \cdot |\omega_o| + \omega_2 \cdot |\omega_2|)$
From: $E_2 + E_0 = \frac{1}{2}.J.\omega_2{}^2 + \frac{1}{2}.J.\omega_0{}^2 = \frac{1}{2}.J.(\omega_2{}^2 + \omega_0{}^2) = E_1 - E_3$ & $\omega^2 = \omega.|\omega|$

$\Delta = \sqrt{[\,5.J / 2.m\,]} / r$

Where the above variables are as defined in Chapters 4.1 to 4.4

$E_0 = \frac{1}{2}.J.\omega_o.|\omega_o|$

$E_2 = \frac{1}{2}.J.\omega_2.|\omega_2|$

$\omega_1 = \text{sign}(E_1) \cdot \sqrt{[\,2.|E_1| / J\,]}$

$\omega_3 = \text{sign}(E_3) \cdot \sqrt{[\,2.|E_3| / J\,]}$

As can be seen from Tables 4.5.1 to 4.5.4, the results are exactly the same as those provided in Tables 4.1.1 to 4.4.1

As can be seen from Table 4.5.5, the results for polar moment of inertia are also exactly the same as those provided in Table 3.3.1

The Theory of Spin

Celestial Body	E_0 (J)	ω_0 (c/s)
The Sun	1.461276E+16	8.65661E-16
Mercury	1.774469E+23	8.26678E-07
Venus	1.733075E+24	3.23644E-07
Earth	1.733075E+24	1.99099E-07
Mars	8.871154E+16	1.05859E-07
Jupiter	2.712883E+23	1.67849E-08
Saturn	3.481068E+21	6.75905E-09
Uranus	3.900860E+19	2.36992E-09
Neptune	7.714895E+19	1.20823E-09
Pluto	1.768504E+17	8.03026E-10

Table 4.5.1

Celestial Body	E_1 (J)	ω_1 (c/s)
The Sun	2.300141E+16	1.086064E-15
Mercury	5.765629E+23	1.490135E-06
Venus	2.515333E+23	1.232980E-07
Earth	3.200840E+23	2.433673E-07
Mars	1.556136E+22	4.433650E-05
Jupiter	2.528419E+26	5.124219E-07
Saturn	9.191740E+24	3.473221E-07
Uranus	2.807922E+22	6.358382E-08
Neptune	2.093597E+21	6.294054E-09
Pluto	6.271063E+15	1.512158E-10

Table 4.5.2

Celestial Body	E_2 (J)	ω_2 (c/s)
The Sun	1.600977E+35	2.865329E-06
Mercury	3.991161E+23	1.239801E-06
Venus	-1.481541E+24	-2.992369E-07
Earth	2.877018E+28	7.296281E-05
Mars	3.977416E+22	7.088236E-05
Jupiter	2.977768E+31	1.758525E-04
Saturn	2.044045E+30	1.637867E-04
Uranus	-7.118296E+28	-1.012377E-04
Neptune	6.202913E+29	1.083383E-04
Pluto	-3.555149E+25	-1.138559E-05

Table 4.5.3

The Theory of Spin

Celestial Body	E_3 (J)	ω_3 (°/s)
The Sun	-1.600977E+35	-2.865329E-06
Mercury	0	0
Venus	0	0
Earth	-2.877008E+28	-7.296268E-05
Mars	-2.421289E+22	-5.530457E-05
Jupiter	-2.977743E+31	-1.758518E-04
Saturn	-2.044077E+30	-1.637864E-04
Uranus	7.118298E+28	1.012377E-04
Neptune	-6.202913E+29	-1.083382E-04
Pluto	3.555149E+25	1.138559E-05

Table 4.5.4

Celestial Body	Δ	J (kg.m²)
The Sun	0.318281959	3.900014E+46
Mercury	0.812862196	5.193084E+35
Venus	0.68123191	3.309127E+37
Earth	0.334460097	1.080860E+37
Mars	0.002317087	1.583269E+31
Jupiter	0.02278067	1.925856E+39
Saturn	0.014060011	1.523954E+38
Uranus	0.024937619	1.389062E+37
Neptune	0.065237927	1.056968E+38
Pluto	8.64241985	5.484999E+35

Table 4.5.5

The Theory of Spin

4.6 Core-Mantle Rotation

The potential and kinetic energies working in this situation act differently. Potential energies act at a satellite's core and kinetic energies act throughout a body's mass.

δω is the relative angular rotation rate between a satellite's core and its mantle.

E_ω is the energy generating δω

The following calculations have been carried out for all of the principal bodies in our solar system using the formulas listed in Chapter 4.6.1:

It should be noted that the dimensions for the gas planet are as we see them. The outside radius, and therefore the densities, of these planets are incorrect in Tables 4.6.3 to 4.6.5 because they are based upon the atmospheric radius, not the planetary radius.

Note: The calculation for Mars has failed (J_m < 0) because it has lost its core; it is hollow. This revelation is not unexpected given its 'Δ' value (0.002317087) despite it not being a gas planet.

		Our Sun	**Mercury**	
Core:	ρ_c	7870	6432.2033746	kg/m³
	r_c	1.44572E+08	458566.0805	m
	m_c	9.9612985E+28	2.5980898E+21	kg
	Δ_c	1	1	
	J_c	8.3280698E+44	2.1853349E+32	kg.m²
	E_c	1.461276E+16	1.7744686E+23	J
	ω_c	5.9239168E-15	4.02986184E-05	°/s
Mantle:	ρ_m	1351.2864468	5420.292608	kg/m³
	r_m	695710000		m
	m_m	1.888887E+30	3.275120E+23	kg
	Δ_m	0.3302708102	0.8307143289	
	J_m	3.816733E+46	5.190899E+35	kg.m²
	E_m	-1.600977E+35	-5.765629E+23	J
	ω_m	-2.896421E-06	-1.490449E-06	°/s
Effective:	δω	**-2.896421E-06**	**3.880817E-05**	°/s
	E_ω	**-1.600977E+35**	**-7.540098E+23**	J
Table 4.6.1				

		Venus	**Earth**	
Core:	ρ_c	6532.59060108	7870	kg/m³
	r_c	1098857.372	1215000	m
	m_c	3.63037E+22	5.91279E+22	kg
	Δ_c	1	1	
	J_c	1.7536409E+34	3.4914411E+34	kg.m²
	E_c	1.73307475E+24	2.1422838E+23	J
	ω_c	1.40589637E-05	3.50308874E-06	c/s
Mantle:	ρ_m	5234.89570932	5492.91159250	kg/m³
	r_m	6051800	6371000.685	m
	m_m	4.8310624E+24	5.8987352E+24	kg
	Δ_m	0.6951614327	0.3418703471	
	J_m	3.307370E+37	1.077369E+37	kg.m²
	E_m	-2.515333E+23	-2.87704E+28	J
	ω_m	-1.233307E-07	-7.308122E-05	c/s
Effective:	$\delta\omega$	**1.393563E-05**	**-6.957813E-05**	c/s
	E_ω	**-1.984608E+24**	**-2.877061E+28**	J
Table 4.6.2				

		Mars	**Jupiter**	
Core:	ρ_c	7870	7870	kg/m³
	r_c	461831.5834	3211180.751	m
	m_c	3.24724E+21	1.09159E+24	kg
	Δ_c	1	1	
	J_c	2.77039E+32	4.50243E+36	kg.m²
	E_c	N/A	2.71288E+23	J
	ω_c	N/A	3.47142E-07	c/s
Mantle:	ρ_m	3924.0995	1325.58260832	kg/m³
	r_m	N/A		m
	m_m	N/A	1.8971E+27	kg
	Δ_m	N/A	0.022784618	
	J_m	**-2.61207E+32**	1.92135E+39	kg.m²
	E_m	N/A	-2.97777E+31	J
	ω_m	N/A	-0.000176058	c/s
Effective:	$\delta\omega$	N/A	**-1.757113E-04**	c/s
	E_ω	N/A	**-2.977768E+31**	J
Table 4.6.3				

The Theory of Spin

		Saturn	**Uranus**	
Core:	ρ_c	7870	7870	kg/m³
	r_c	1385801.125	1115922.071	m
	m_c	8.77336E+22	4.58105E+22	kg
	Δ_c	1	1	
	J_c	6.7395E+34	2.28188E+34	kg.m²
	E_c	3.481068E+21	3.90086E+19	J
	ω_c	3.21405E-07	5.84721E-08	c/s
Mantle:	ρ_m	687.0262033	1269.8529212	kg/m³
	r_m	58232000	25362000	m
	m_m	5.68252E+26	8.67672E+25	kg
	Δ_m	0.014061969	0.024947866	
	J_m	1.52325E+38	1.38678E+37	kg.m²
	E_m	-2.044087E+30	7.118296E+28	J
	ω_m	-1.638230E-04	1.013209E-04	c/s
Effective:	$\delta\omega$	**-1.635016E-04**	**1.013794E-04**	c/s
	E_ω	**-2.044045E+30**	**7.118296E+28**	J
Table 4.6.4				

		Neptune	**Pluto**	
Core:	ρ_c	7870	7870	kg/m³
	r_c	1396768.826	76464.36337	m
	m_c	8.98332E+22	1.47381E+19	kg
	Δ_c	1	1	
	J_c	7.01045E+34	3.44682E+28	kg.m²
	E_c	7.7148954E+19	1.76850379E+17	J
	ω_c	4.691449E-08	3.2033813E-06	c/s
Mantle:	ρ_m	1636.7964485	1858.3531833	kg/m³
	r_m	24622000	1187000	m
	m_m	1.02323E+26	1.30153E+22	kg
	Δ_m	0.065350148	8.665309303	
	J_m	1.05627E+38	5.485E+35	kg.m²
	E_m	-6.202913E+29	3.55515E+25	J
	ω_m	-1.083742E-04	1.138559E-05	c/s
Effective:	$\delta\omega$	**-1.083273E-04**	**1.458897E-05**	c/s
	E_ω	**-6.202913E+29**	**3.555149E+25**	J
Table 4.6.5				

The Theory of Spin

As all planets (and stars) originally comprised similar matter immediately after the 'Big-Bang', it is possible to estimate the core-mantle properties of the gas planets based upon the expected densities.

Assuming the largest planets (including the earth) all have similar percentages of iron content, it is also reasonable to assume that a gas planet, with very mobile matter, will have allowed a similar percentage of its iron to have migrated to its core. Therefore, it is also possible to estimate the core radius of the gas planets based upon the earth's properties:

$$r_c/r_{cE} = r_m/r_{mE} \cdot \rho_p/\rho_E$$

Where:

r_c = the radius of the core of the gas planet

r_{cE} = the radius of the earth's core

r_m = the outside radius of the gas planet

r_{mE} = the outside radius of the earth

ρ_p = the density of the gas planet (5392 kg/m³)

ρ_E = the density of the earth

Using the above formula, it has been possible to estimate the angular rotation rate of the surface of the planets (ω_m) along with their core-mantle relative rotations ($\delta\omega$). Their internal [heat] energies (E), will of course remain unchanged.

The following Tables (4.6.6 & 4.6.7) list the core-mantle properties of the gas planets assuming they comprise similar matter to that of the three innermost planets (5392 kg/m³).

As expected, the energies (E_ω) remain unchanged but the relative spin-rates ($\delta\omega$) have increased as follows:

Jupiter: 15.6%; Saturn: 32.1%; Uranus: 11.9%; Neptune: 1.7%

Indicating that the actual planetary densities are slightly different to those estimated (5392 kg/m³).

		Jupiter	Saturn	
Core:	ρ_c	7870	7870	kg/m^3
	r_c	8181077.811	5473099.840	m
	m_c	1.805074E+25	5.404599E+24	kg
	Δ_c	1	1	
	J_c	4.832545E+38	6.475752E+37	kg.m^2
	E_c	2.7128826E+23	3.4810679E+21	J
	ω_c	3.350754E-08	1.036874E-08	c/s
Mantle:	ρ_m	5376.787354	5376.787354	kg/m^3
	r_m	43799850.538	29301879.348	m
	m_m	1.880139E+27	5.629354E+26	kg
	Δ_m	0.036361300	0.027941925	
	J_m	1.9258557E+39	1.5239539E+38	kg.m^2
	E_m	-2.977768E+31	-2.044087E+30	J
	ω_m	-2.031830E-04	-2.159826E-04	c/s
Effective:	$\delta\omega$	**-2.031495E-04**	**-2.159722E-04**	c/s
	E_ω	**-2.9777681E+31**	**-2.0440866E+30**	J

Table 4.6.6

		Uranus	Neptune	
Core:	ρ_c	7870	7870	kg/m^3
	r_c	2925671.432	3091355.794	m
	m_c	8.255436E+23	9.738909E+23	kg
	Δ_c	1	1	
	J_c	2.826514E+36	3.722788E+36	kg.m^2
	E_c	3.9008604E+19	7.7148954E+19	J
	ω_c	5.253753E-09	6.437924E-09	c/s
Mantle:	ρ_m	5376.787354	5376.787354	kg/m^3
	r_m	15663458.338	16550499.198	m
	m_m	8.598746E+25	1.014391E+26	kg
	Δ_m	0.040378561	0.097053764	
	J_m	1.3890623E+37	1.0569685E+38	kg.m^2
	E_m	7.118296E+28	-6.202913E+29	J
	ω_m	1.134344E-04	-1.102981E-04	c/s
Effective:	$\delta\omega$	**1.134397E-04**	**-1.102916E-04**	c/s
	E_ω	**7.1182956E+28**	**-6.2029129E+29**	J

Table 4.6.7

4.6.1 Formulas

The following constitutes the calculation sequence to establish the relative rotation rate of a planet's (or star's) core and its mantle:

Input Data

ρ_c = 7870	{kg/m³}	(iron core)
r_c = 1215000	{m}	(e.g. Earth)
ρ_m = 5506.351327	{kg/m³}	(e.g. Earth)
r_m = 6371000.685	{m}	(e.g. Earth)
$E_c = E_0$	{J}	
$\Delta_c = 1$		(solid core of constant density)

Output Data

Core:

$m_c = \rho_c / (4/3.\pi.r_c^3)$ {kg}

$J_c = \tfrac{2}{5}.m_c.r_c^2$ {kg.m²}

$\omega_c = \sqrt{[\,2.E_c / J_c\,]}$ {c/s}

Mantle:

$m_m = \rho_m / (4/3.\pi.r_m^3)$ {kg}

$J_m = J - J_c$ {kg.m²}

$\Delta_m = \sqrt{[\,5/2.J_m / m_m.(r_m^2 - r_c^2)\,]}$

$E_m = E_3 - E_1$ {J}

$\omega_m = \sqrt{[\,2.E_m / J_m\,]}$ {c/s}

Effective:

$\delta\omega = \omega_m - \omega_c$ {c/s}

$E_\omega = E_m - E_c$ {J}

4.7 No moon

If it lost its moon, the earth's 'E_3' would be zero and its rotational energy would become:

$E_2 = E_1 - E_0 - 0 = 3.207996E+23 - 2.144732E+23 = 1.063264E+23$ {J}

$E_2 = ½.J.\omega_o^2 \qquad \rightarrow \qquad \omega_o = \sqrt{[\, 2.E_2/J \,]} = 1.401854E-07$ {c/s}

$t_s = 2\pi/\omega_o = 4.482055E+07$ {s}

or 12450.1516 hours (<519 days).

Moreover, the earth would lose 99.99963% of its internal [heat] energy; i.e. its surface temperature would fall by approximately 210 K: no more continental drift.

We can therefore conclude that the presence of our moon is essential for life on earth.

4.8 Magnetic Fields

Spin in a satellite's iron core is dominated by its force-centre and spin in its iron-rich mantle is dominated by its sub-satellite(s). The relative rotation of a satellite's electrical and magnetic charges (e' & m) in its mantle and core are together responsible for generating its polar magnetic field.

The fact that only non-hollow planets (i.e. excluding Mars) with moons (i.e. excluding Mercury & Venus), actually generate a magnetic field normal to the orbital plane of their satellites, and those that generate the greatest spin-energy (E_ω) also generate the greatest magnetic field is additional proof of spin theory.

As I have set prograde direction as negative (E_3 is negative for the earth), and because '$\delta\omega$' is also negative, the earth's mantle (including its crust) must be spinning in a prograde direction. Using the right-hand rule, the magnetic north pole of the earth's should be pointing in the direction of our North Pole; which it is.

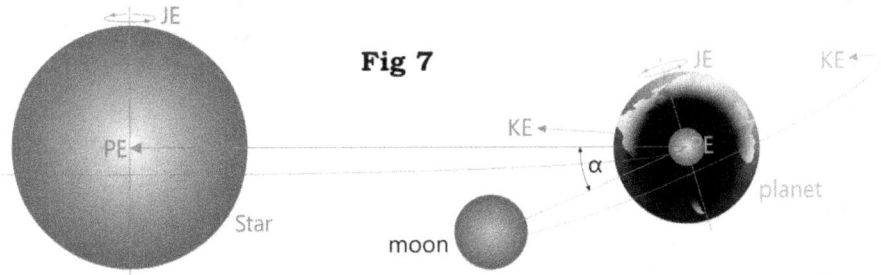

Fig 7

The earth's lunar tilt angle (Fig 7: α = 23.4°) is a clear indication that the earth acquired its moon from outside its solar system (galactic comets), as is the case for all of the moons in our solar system, and perhaps, most of its planets.

If a planet's lunar orbital plane is not coincident with its solar orbital plane, its mantle and core will spin on different axes, generating an angular difference between its true (physical) North and its magnetic North.

The Theory of Spin

4.8.1 Formulas

There are two competing energies driving the spin in the earth's core and its mantle:
- $-E_0$ (-2.1447324E+23 J) is the sun's energy driving the core
- E_1-E_3 (2.87704E+28 J) is the moon's energy driving the mantle (and core)

The polar moments of inertia:
Core: J = 3.49144112166E+34 kg.m²
Mantle: J_m = 1.07860404E+37 kg.m²

The angular velocities:
Core: ω = Sign(E_0) . $\sqrt{[2.|E_0|/J]}$ = -3.50509019131E-06 ᶜ/s
Mantle: ω_m = Sign(E_1-E_3) . $\sqrt{[2.|E_1-E_3|/J_m]}$ = 7.3039350764E-05 ᶜ/s

The differential angular velocity:
$\delta\omega = \omega + \omega_m$ = 6.95342605725E-05 ᶜ/s

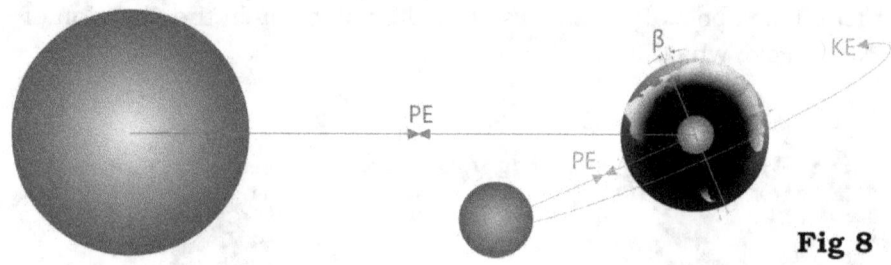

Fig 8

The angular tilt (β) between the two axes can be calculated thus:
$\beta = \text{sign}(\omega/\omega_m) . \frac{1}{2}\sqrt{[|Asin(\omega/\omega_m)|]}$
 = 0.109553685228394 radians (**6.27696379369167°**)

4.9 Magnetic Reversal

A worrying aspect of the earth's magnetic field is that the relative spin induced in the earth and its core by our moon and our sun will not reverse unless either the earth or its moon changes orbital direction, which is highly unlikely. Something external to the earth (extra-terrestrial) must therefore cause this reversal to occur periodically.

It would appear that the earth's magnetic reversal can only be explained by flipping it through 180°; switching north and south poles!

We already know that our solar system has a number of orbiting comets, so it is highly likely that the Milky Way also has its own 'comets', and these could be planet sized. Therefore, a large galactic comet may well be responsible for flipping the earth and/or any other planet in the solar system as it passes close by.

The Theory of Spin

5 Atomic Spin

The same spin theory applies to the proton-electron pair as to celestial bodies.

However, protons and electrons both have consistent densities, and once paired, a proton will collect no more electron partners, limiting their spin performance.

This spin is present in an electron after being forcibly removed from its orbit, and will remain with it until it is once again paired to a lone proton. The impact between spinning electrons, whether free-flying or orbiting, is what we see in bubble chambers.

When a proton-electron pair combine to create a neutron, the outer surface of the proton is revolving at a curvilinear speed 25.7 times slower than that of the orbiting electron.

Hypothesis: *This velocity variation creates an internal 'stress' within a neutron that ultimately leads to its disintegration into alpha and beta particles. Neutron-neutron interaction inside an atomic nucleus will aggravate this stress reducing its disintegration period:* **half-life**

5.1 Formulas

$E_0 = \frac{1}{2}.J.\omega_0|\omega_0|$

Where:
$\omega_0 = 2.\pi/t_0 = 2\pi/t_n . (T/T_n)^{1.5}$
t_0 = electron orbital period
T = electron temperature
t_n = neutronic period (5.90596121302193E-23 s)
T_n = neutronic temperature (623316124.717178 K)
$J = \frac{2}{5}.m.r^2 = 7.66586456056651\text{E-}63$ kg.m^2
$m_e = 9.1093897\text{E-}31$ kg
$r = 1.45046059426276\text{E-}16$ m

$E_1 = \delta KE . (r/R)^2$

as the velocity of an orbital electron doesn't vary within its orbit at any given temperature, $\delta KE = 0$

$E_3 = \text{sign}(\cos(\theta)).(\Sigma KE^P + \Sigma PE^A)$

$KE^P = \frac{1}{2}.m_e.v^2 = \frac{1}{2}.m_e.T/X$

Where:
T = electron temperature
$m_e = 9.1093897\text{E-}31$ kg
$X = 6.9353271647894\text{E-}09$ K.s^2/m^2

$E_2 = E_1 - E_0 - E_3$

The Theory of Spin

At the neutronic temperature: T_n = 623316124.717178 K

$E_2 = E_3$ = 4.09789381263211E-14 J (using above formulas)
Which is the same as: KE = ½.m_e.c^2 = 4.09789381263211E-14 J

The total energy is therefore:
E = PE + KE + E_2
E = m_e.c^2 + ½.m_e.c^2 + E_2 = 0

Where:
c = 299792459 m/s
m_e = 9.1093897E-31 kg

i.e.; all the forces balance.

The spin rate of an orbiting electron at 'c' is:
ω_o = 2.π/t_n = 1.06387175271756E+23 c/s

The spin rate of a proton when its electron is orbiting at 'c' is:
ω_p = √[2.E_3/J_p] = 6.2281789631761E+21 c/s
and its surface is travelling at a curvilinear speed of:
v = ω_p.r_p = 11062072.3389274 m/s
Where: r_p = 1.77613270336827E-15 m
The velocity ratio is:
δ_v = v/c = 11062072.3389274 / 299792459
δ_v = 0.036899101384426

The electron will be orbiting at the neutronic radius:
R_n = 2.81793795383896E-15 m

The combined radii of a proton and a neutron is:
R = r_p + r_e = 1.45046059426276E-16 + 1.77613270336827E-15
R = 1.92117876279455E-15 m

The radial ratio is:
δ_R = R_n/R = 2.81793795383896E-15 / 1.92117876279455E-15
δ_R = 1.46677550700175

The Theory of Spin

6 Spin in Our Galaxy & Solar System

The impact of spin will differ according to where, and on what, it is acting.

As all the matter in the universe originated from the same place (the ultimate body), it will be similar. However, during any universal period, the internal frictional heat generated by celestial spin will result in celestial bodies that comprise matter that varies from a maximum at their core (≈ 8000 kg/m^3) to a minimum immediately below their crust (≈ 1000 kg/m^3).

When celestial bodies are destroyed through impact, this variation in matter is responsible for creating satellites of different densities.

The majority of the heat within a satellite with sub-satellites of its own, is generated through internal spin friction. The larger the satellite and the greater its sub-satellite population mass, the more heat energy it will generate, hold and radiate. Radiated energy from its force-centre (e.g. star) is responsible for only a small amount of a satellite's heat energy.

The only reason gas planets can exist (as gas planets), so far from their force-centre (e.g. Neptune) is because all of its body heat is generated by planetary spin.

We know that life exists in the dark (caves and deep ocean) and in extreme heat, subsea fumaroles, so there is no reason to assume that some form of life cannot exist somewhere on a gas planet.

Whilst it is currently hypothesis, it is likely that spin is responsible for the half-life of atoms; neutron energy-stress. If so, it is reasonable to claim that spin is as important for the workings of the universe as orbits.

6.1 Celestial Spin

Spin in a celestial body generates internal friction. The energy to generate this spin comes from celestial orbits.

Spin will not occur in any celestial body that is not itself orbiting a force-centre. That is why all galactic force-centres are dark; they possess no internal heat.

All orbiting satellites – except those in a linear orbit - possess natural spin (ω_o).

The kinetic spin energy generated in a force-centre by its satellites will counter its satellite's natural spin (ω_o), but its effect will be minimal: $\delta\omega \approx 0$

After a non-linear satellite has acquired sub-satellites of its own, it will begin to generate significant internal heat through friction. The greater its sub-satellite population mass, the greater its internal heat.

Spin will be negligible in a satellite that has no satellites of its own. That is why all moons are cold; they generate no internal heat. However, whilst a moon orbiting sufficiently close to a large planet - and therefore quickly - will generate some internal heat, it will not be through spin, but through varying gravitational action on its core.

Planets with no moons, such as Mercury and Venus, will generate some internal heat if their force-centre's spin period is different to the planet's orbital period.

Highly active planets, i.e. those with significantly active moons (e.g. Mars), can actually suffer considerable disruption through spin.

If a celestial body is sufficiently large, and has acquired a sufficient sub-satellite population, it will eventually generate and hold on to, the heat energy required to achieve the neutronic temperature (T_n) in its core. After which, fissionable decay will occur. This is when a celestial body becomes bright (e.g. a star).

Every bright star in the night-sky *must* host a planetary population.

The Theory of Spin

The average density of the inner planetary satellites is about 5392 kg/m³

The nearest matter density to this is Germanium (5323 kg/m³) and its specific heat capacity is 384 J/kg/K

Therefore, the temperature generated in each body as a result of internal frictional heat can be calculated thus: $\underline{T}$ = E / m.SHC

The following Table lists the temperature generated in the bodies in our solar system:

Body	Mass (kg)	Core energy (J)	Temperature (K)
Sun	1.9885E+30	1.601005E+35	209.67
Mercury	3.3011E+23	7.540098E+23	0.006
Venus	4.86737E+24	1.984608E+24	0.001
Earth	5.96452E+24	2.877018E+28	12.562
Mars	6.4171E+23	N/A	0
Jupiter	1.89819E+27	2.977768E+31	40.853
Saturn	5.6834E+26	2.044045E+30	9.366
Uranus	8.6813E+25	7.170495E+28	2.151
Neptune	1.02413E+26	6.119540E+29	15.561
Pluto	1.303E+22	3.555149E+25	7.105
Table 6.1.1			

Whilst the above temperatures are being generated continually, once a balance is reached between the energy held within the body and that radiated, the body's surface temperature will stabilise.

The above Table reveals quite a picture, which is explained in the following Chapters.

The Theory of Spin

6.1.1 Hades

Hades is in a linear orbit. It is travelling in a straight-line away from its point of origin at the beginning of this universal period; the '*Big-Bang*'.

mass	1.7657202E+41	kg
average density	9000	kg/m³
Δ	1	
surface spin rate ω	-6.65487187E-08	c/s
core-mantle spin rate δω	0	c/s
core-mantle spin energy (E$_3$)	-4.3781616E+50	J
spin energy per unit mass	N/A	J/kg
Temperature generated	0	K
Table 6.1.1		

Hades spins because its satellites (stars) cause it to do so. The number of galactic satellites it hosts will determine its spin rate. The above value (ω) is based upon my own guess at the Milky Way's star-system population of 10bn. NASA's guess is 100bn, which if correct would cause Hades to spin at 6.655E-08 radians per second.

Because Hades' core-mantle spin rate is zero, it generates no internal frictional heat; it is cold. That is why we cannot see it. It radiates negligible electro-magnetic energy.

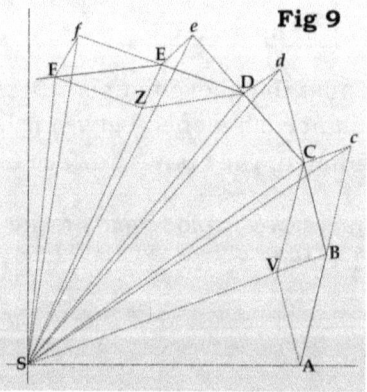

Fig 9

But we know it exists, because Newton told us so (Fig 9; 'S'):
His fundamental law of orbital motion requires that every orbital system *must* include '<u>only one force-centre</u>' and at least <u>one satellite</u>.

6.1.2 Sun

Spin Energy is responsible for most of the sun's heat. But because the matter in its core has reached the neutronic temperature, it is also generating heat through fission.

	mass	1.9885E+30	kg
	average density	1409.782932	kg/m³
	Δ	0.3182819591	
	surface spin rate ω	2.865329E-06	c/s
	core-mantle spin rate δω	-2.896421E-06	c/s
	core-mantle spin energy	-1.6009772E+35	J
	spin energy per unit mass	80513.1879	J/kg
	Temperature generated	211.873164	K
Table 6.1.2			

The closeness of our sun's core-mantle and surface spin-rates reveals the immense anchoring hold its force-centre (Hades) has on its core, which is therefore responsible for the immense internal heat it generates.

What makes our sun a star, is that the temperature at its core is sufficient to create neutrons. This is the reason its 'Δ' value is so high for what is essentially, a gas satellite. Most of the elements it is creating are neutron-rich and therefore heavy.

The temperature generated in the sun through spin friction is 16.7 times greater than that in the earth (excluding heat radiated by the sun). But the surface temperature of the sun is 27.5 times greater. The difference between these two values is due to the additional heat generated in the sun from energy released through fissionable decay.

Our sun also generates the largest magnetic field in our solar system, which is unsurprising given its size and spin energy, but also confirms that most universal matter is iron-rich.

6.1.3 Mercury

Spin Energy is negligible inside Mercury (2.3 J/kg). The only reason one side of Mercury is hot is due to its proximity to its force-centre's radiated heat energy.

mass	3.3011E+23	kg
average density	5427.012135	kg/m³
Δ	0.812862196	
surface spin rate ω	1.239801E-06	°/s
core-mantle spin rate δω	3.880817E-05	°/s
core-mantle spin energy	-7.5400979E+23	J
spin energy per unit mass	2.2841	J/kg
Temperature generated	0.006	K
Table 6.1.3		

Heat energy from internal friction is generated throughout the life of the planet, whilst in orbit. So, the many billions of years that Mercury has been in orbit around the sun should have allowed its core temperature to build up. But as with all energy, if the surface of the planet is radiating heat faster than it can be collected, its core will lose all of its heat.

This is the case with Mercury, as can be discerned by its 'Δ' value. Almost none of its internal matter is mobile.

6.1.4 Venus

Spin Energy is even less than that inside Mercury (0.41 J/kg).

mass	4.86737E+24	kg
average density	5242.664311	kg/m³
Δ	0.68123191	
surface spin rate ω	-2.992369E-07	°/s
core-mantle spin rate δω	1.3935633E-05	°/s
core-mantle spin energy	-1.984608E+24	J
spin energy per unit mass	0.4077	J/kg
Temperature generated	0.001	K
Table 6.1.4		

However, Venus's 'Δ' value is sufficient to show some internal mobility which means that some of its internal matter will be ejected (in earthquakes and small volcanoes) over time, and is the reason why Venus has an atmosphere. This atmospheric heat is not generated by internal friction, it is generated by the sun's radiated energy. Venus receives 1.9 times as much of the sun's radiated heat energy as that trapped by the earth.

It is currently thought that Venus's surface is flat because of atmospheric pressure. This is incorrect; its flat surface is because the planet has no tectonic plates; its crust is too thick. It has far less volcanic activity than the earth because of its lack of internal spin energy. Whilst it has more than Mercury, it also has much more mass to heat up.

6.1.5 Earth

Before the earth trapped its moon, it was not the raging inferno depicted in countless documentaries and books over recent years, it was cold & dark.

Spin energy generated in the earth by its moon is responsible for its life supporting climatic conditions.

mass	5.95786303764E+24	kg
average density	5509.426596	kg/m³
Δ	0.334277698	
surface spin rate ω	7.292115E-05	°/s
core-mantle spin rate $\delta\omega$	-6.95781266E-05	°/s
core-mantle spin energy	-2.87706106E+28	J
spin energy per unit mass	4823.6258	J/kg
Temperature generated	12.70793446	K
Table 6.1.5		

Its moon is generating over 10,000 times more internal heat energy in the earth than that generated within Venus. It is this energy that drives our mantle plumes, magnetic field and continental drift. It is also responsible for the planet's volcanoes and earthquakes and therefore its weather.

The earth's crust is about 1E+22kg, or < 0.2% of the planet's total mass. An increase in E_3 of a similar margin would be sufficient to melt its crust.

The earth's current life-sustaining energy balance is therefore quite sensitive; by trapping one additional outlying satellite, smaller than its current moon, the earth could become another gas planet.

Therefore, it may be assumed that a generated temperature greater than ≈15 K is likely to result in a gas planet. But this value only applies to a planet density similar to that of the earth. A lower generated temperature would be needed to achieve such an event in a planet of lower density.

78% of the earth's atmosphere (nitrogen) has been generated over the last 0.5bn years by its animal-life.

21% of the earth's atmosphere (oxygen) has been generated over the last 2bn years by its plant-life (stromatolites).

Almost 1% of its atmosphere (argon) has been generated by potassium decay during the planet's lifetime.

The Theory of Spin

Less than 0.04% of its atmosphere is created by the earth itself from continental drift.

Therefore, during its early life, earth's atmosphere must have been very thin indeed; <1% of today's mass and argon-rich.

And yet there was liquid water on its surface more than 4bn years ago[#]

[#] This anomaly could be explained by the presence of its moon (more than 4bn years ago), that together with the sun's electro-magnetic radiation, generated sufficient surface heat to keep its water above its triple-point, which, considering the low atmospheric pressure, could have existed in liquid form below 273K

The obvious conclusion to the above is that we are far more likely to find solid rock and gas planets in our universe than those with such a fine balance as in our Earth.

The Theory of Spin

6.1.6 Mars

Mars is a fascinating planet. Small though it is, its largest moon (Phobos) is incredibly active. Its density appears to indicate that Mars is now mostly hollow.

mass	6.4171E+23	kg
average density	3934.080869	kg/m³
Δ	0.002317087	
surface spin rate ω	7.088236E-05	°/s
core-mantle spin rate $\delta\omega$	N/A	°/s
core-mantle spin energy	N/A	J
spin energy per unit mass	N/A	J/kg
Temperature generated	N/A	K
Table 6.1.6		

The core mantle energy calculation failed because Mars has lost its core. *Note its 'Δ' value.* This is the reason it *looks* dead. However, ...

There must be a very good reason why our smallest planet has created the solar system's largest volcanoes. Given Phobos's activity and the fact that such a small planet has two moons (incl. Deimos), means that it must have had a very active core at some time in its past.

It is highly likely that whilst it hosted just one moon, it was a smaller planet with surface water, sufficient activity in its core to generate the heat necessary to create oxygen emitting plant-life.

When it acquired a second moon, core activity became too much for the planet to contain. It blasted its inner (iron-rich) matter onto its surface through massive volcanoes, where the oxygen became locked in the iron. Hence its rust colour.

Its surface water subsequently seeped into the planet's voids inside where it now resides.

Put all of this together and you end up with a hollow planet into which its water has soaked, perhaps together with its surface life and also its atmosphere. Now that would be an interesting place to explore. Especially if the energy from its moons is sufficient to keep its internal water liquid.

The Theory of Spin

6.1.7 Asteroid-Belt (Ceres)

None of the rubble orbiting in the asteroid belt have moons of their own.

mass	9.47E+20	kg
average density		kg/m³
Δ		
surface spin rate ω		°/s
core-mantle spin rate δω		°/s
core-mantle spin energy		J
spin energy per unit mass	N/A	J/kg
Temperature generated		K
Table 6.1.7		

Therefore, spin plays no part in the life of this rubble.

However, it does influence the spin in its force-centre; the sun, albeit only minimally.

6.1.8 Jupiter

Jupiter is a gas planet. A gas planet is one that has collected sufficient sub-satellite population to generate the internal heat to melt its crust.

mass	1.89819E+27	kg
average density	1326.216812	kg/m³
Δ	0.02278067	
surface spin rate ω	1.758525E-04	°/s
core-mantle spin rate $\delta\omega$	-1.7571129E-04	°/s
core-mantle spin energy	-2.977768E+31	J
spin energy per unit mass	15687.4058	J/kg
Temperature generated	41.28265299	K

Table 6.1.8

The dimensions used to calculate the planet's 'Δ' value are based upon the planet we can see, which includes its atmosphere.

If we assume that like all other large planets, excluding its atmosphere, Jupiter is iron-rich with a density of 5392 kg/m³, the above Table would look just the same but only if the *planet* was spinning at the same angular velocity as its atmosphere. Given the prograde-retrograde mixture of its satellite orbits, this is unlikely.

If its 'Δ' value is set to a more representative '0.3', i.e. slightly less than that for the earth, the *planet* Jupiter must be spinning at an angular velocity of 2.13E-05 °/s; i.e. more than eight times slower than its atmosphere.

This would also mean that *planet* Jupiter is dimensionally almost seven times greater than the earth and spinning at less than one third of the earth's angular velocity.

The temperature generated by its moons is more than enough to turn Jupiter into a gas planet assuming a similar density to that of the earth (refer to Chapter 6.1.5)

The Theory of Spin

6.1.9 Saturn

Saturn is a gas planet. A gas planet is one that has collected sufficient sub-satellite population to generate the internal heat to melt its crust.

mass	5.6834E+26	kg
average density	687.1230137	kg/m³
Δ	0.014060011	
surface spin rate ω	1.637867E-04	c/s
core-mantle spin rate δω	-1.63501557E-04	c/s
core-mantle spin energy	-2.04408656E+30	J
spin energy per unit mass	3596.5176	J/kg
Temperature generated	9.46471331	K
Table 6.1.9		

The dimensions used to calculate the planet's 'Δ' value are based upon the planet we can see, which includes its atmosphere.

If we assume that like all other large planets, excluding its atmosphere, Saturn is iron-rich with a density of 5392 kg/m³, the above Table would look just the same but only if the *planet* was spinning at the same angular velocity as its atmosphere. Given the prograde-retrograde mixture of its satellite orbits, this is unlikely.

If its 'Δ' value is set to a more representative '0.3', i.e. slightly less than that for the earth, the *planet* Saturn must be spinning at an angular velocity of 1.525E-05 c/s; i.e. almost eleven times slower than its atmosphere.

This would also mean that *planet* Saturn is dimensionally 4.6 times greater than the earth and spinning at just over one fifth of the earth's angular velocity.

The generated temperature (Table 6.1.9), however, would indicate that Saturn's density is actually less than that used in the above calculation: a lower density of the same mass would result in greater energy generation and a lower gas/vapour point.

6.1.10 Uranus

Uranus is a gas planet. A gas planet is one that has collected sufficient sub-satellite population to generate the internal heat to melt its crust.

mass	8.6813E+25	kg
average density	1270.415139	kg/m³
Δ	0.024937619	
surface spin rate ω	-1.012377E-04	°/s
core-mantle spin rate δω	1.013794E-04	°/s
core-mantle spin energy	7.11829563E+28	J
spin energy per unit mass	825.9702	J/kg
Temperature generated	2.157782466	K
Table 6.1.10		

The dimensions used to calculate the planet's 'Δ' value are based upon the planet we can see, which includes its atmosphere.

If we assume that like all other large planets, excluding its atmosphere, Uranus is iron-rich with a density of 5392 kg/m³, the above Table would look just the same but only if the *planet* was spinning at the same angular velocity as its atmosphere. Given the prograde-retrograde mixture of its satellite orbits, this is unlikely.

If its 'Δ' value is set to a more representative '0.3', i.e. slightly less than that for the earth, the *planet* Uranus must be spinning at an angular velocity of -1.368E-05 °/s; i.e. almost seven and a half times slower than its atmosphere.

This would also mean that *planet* Uranus is dimensionally almost two and a half times greater than the earth and spinning at almost one fifth of the earth's angular velocity.

The generated temperature (Table 6.1.10), however, would indicate that Uranus's density is considerably less than that used in the above calculation: a lower density of the same mass would result in greater energy generation and a lower gas/vapour point. It would appear that in Uranus's case, most of its matter is ice and/or low-density gases in viscous form.

6.1.11 Neptune

Neptune is a gas planet. A gas planet is one that has collected sufficient sub-satellite population to generate the internal heat to melt its crust.

mass	1.02413E+26	kg
average density	1637.934377	kg/m³
Δ	0.065237927	
surface spin rate ω	1.083382E-04	°/s
core-mantle spin rate δω	-1.0832728E-04	°/s
core-mantle spin energy	-6.20291286E+29	J
spin energy per unit mass	5975.3544	J/kg
Temperature generated	15.93885045	K
Table 6.1.11		

The dimensions used to calculate the planet's 'Δ' value are based upon the planet we can see, which includes its atmosphere.

If we assume that like all other large planets, excluding its atmosphere, Neptune is iron-rich with a density of 5392 kg/m³, the above Table would look just the same but only if the *planet* was spinning at the same angular velocity as its atmosphere. Given the prograde-retrograde mixture of its satellite orbits, this is unlikely.

If its 'Δ' value is set to a more representative '0.3', i.e. slightly less than that for the earth, the *planet* Neptune must be spinning at an angular velocity of -3.481E-05 °/s; i.e. three times slower than its atmosphere.

This would also mean that *planet* Neptune is dimensionally 2.6 times greater than the earth and spinning at less than one half of the earth's angular velocity.

The temperature generated by its moons is enough to turn Neptune into a gas planet assuming a similar density to that of the earth (refer to Chapter 6.1.5).

6.1.12 Pluto

Pluto has acquired so many comparatively large moons that it is being pulled into a local orbit, which is the reason for its 'Δ' value greater than 1

mass	1.303E+22	kg
average density	1859.960193	kg/m³
Δ	8.64241985	
surface spin rate ω	-1.138559E-05	°/s
core-mantle spin rate δω #	1.45889736E-05	°/s
core-mantle spin energy #	3.55514894E+25	J
spin energy per unit mass #	2728.4336	J/kg
Temperature generated #	7.18008834	K
Table 6.1.12		

These values are unreal. They would only be valid for a planet that was spinning about its centre of mass.

As can be seen from the above Table, if this were the case, it would be generating more than three times the temperature generated by Uranus and more than half that generated by the earth.

However, because it is being pulled into a localised orbit, the sun's potential energy (E_1) is not acting within the planet. Therefore, it cannot have an active core. But this does not detract from the fact that Pluto is a very active planet.

The Theory of Spin

6.1.13 Moon

Our moon is our only satellite and it has no sub-satellites of its own. Therefore, it is driven only by ω_o (2.66167E-06 c/s), which is the same as its orbital angular velocity.

mass	7.346377E+22	kg
average density	3343.599878	kg/m^3
Δ	N/A	
surface spin rate ω	N/A	c/s
core-mantle spin rate $\delta\omega$	N/A	c/s
core-mantle spin energy	N/A	J
spin energy per unit mass	N/A	J/kg
Temperature generated	N/A	K
Table 6.1.13		

Because its E_1 energy is negligible and E_3 energy is zero, our moon is not going to generate any internal heat energy. It is therefore inactive.

6.2 Atomic Spin

The proton-electron pair is where atomic spin is generated, and it occurs in both particles.

Spin does not generate heat inside a proton because only E_3 is present in the proton-electron partnership; i.e. there are no competing spin energies.

An orbiting electron's spin period is the same as its orbital period.

The spin in a proton will always be less than that in its electron. The spin difference reduces with increasing temperature.

When the pair unite (as a neutron), the electron is spinning more than 17 times faster than its proton partner, and its relative surface speed is 25.7 times greater. It is assumed that the resultant [stored] stress is responsible for a neutron's eventual decay into alpha and beta particles.

If this is the case, atomic spin is responsible for natural radioactivity, and therefore just as important for universal performance as planetary orbits.

We know this to be correct because of the 'Little Boy' atom bomb, 1kg of which released **6.3E+13 J**[#] of energy when it was dropped on Hiroshima.
[#] *empirical value*

The following particle energies are stored at the time of a neutron's creation: $E = |KE_e| + |PE| + |SE| + |KE_p|$
[#] *'electron kinetic energy + potential energy in proton-electron pair + <u>spin energies in electron and proton</u>' at the time of neutron creation + 'repulsion in a proton that has lost its neutrons'*

According to the Newtonian atomic model, the energy contained within a neutron is therefore:

$KE_e = \frac{1}{2}.m_e.c^2 = |\frac{1}{2} \times 9.1093897E{-}31 \times 2997924592|$
  = 4.09355561131267E-14 J

$PE = -m_e.c^2 = |-9.1093897E{-}31 \times 2997924592|$
  = 8.18711122262534E-14 J

$SE = SE_e + SE_p = |-4.09355561131267E{-}14 + 4.33820131944073E{-}17|$
  = 4.09789381263211E-14 J

$KE_p = k.e^2/(4.R_n^2/(R_p+R_e)) = 1.3954267683677E{-}14$ J (proton energy)

The Theory of Spin

Uranium 235 has a neutronic ratio of: ψ = 1.587270761

1 kg of which contains 3.66585231725022E+26 neutrons (N_n) and 2.30953181248165E+26 protons (N_p)

According to Newton's atomic model 'Little Boy' released:

$E = N_n.(KE_e + PE + SE) + N_p.KE_p =$ **6.32642E+13 J** of explosive energy

Confirming;

a) spin energy

b) R_n

c) circular orbits (in atoms)

d) Newton's atomic model

e) the neutronic model

f) neutronic energy

If circular orbits are now confirmed, Quantum Theory must be dead

Because 'R_n' is now confirmed, the true nature of $E=mc^2$ is also confirmed, therefore Relativity must be dead

(because according to Relativity, long before an orbiting electron achieved velocity 'c' it would be orbiting inside the proton - impossible)

The Theory of Spin

Appendices

References, symbols, glossary, etc. used throughout this book along with a summary list of corollaries and hypotheses.

The Theory of Spin

A1 References

Philosophiæ Naturalis Principia Mathematica Rev. IV; Keith Dixon-Roche; 978-1-07215-605-5

The Atom; Keith Dixon-Roche; 978-1-08610-029-7

The Life & Times of the Neutron; Keith Dixon-Roche; 978-1-08251-683-2

The Physical Constants; Keith Dixon-Roche; 978-1-79422-609-8

The Universe; Keith Dixon-Roche; 978- 1-70753-878-2

The Spin Calculator: http://calqlata.com/proddetail.asp?prod=00085

The Theory of Spin

A2 Glossary

Atomic Particle	One of the three components that comprise an atom
Big-Bang	The eruption that occurred when the Ultimate-Body accumulated sufficient 'mass' to compromise the integrity of the innermost neutron.
Black-Body	A collection of Quanta too cold to emit electro-magnetic radiation in the frequencies that would enable detection.
Gas	Atoms that possess greater electrical field energy than magnetic field energy
Hades	The Milky Way's force-centre
Proton-electron pair	A proton that hosts an orbiting electron
Ultimate-Body	A body that contains all the Quanta in the universe ($\approx 2.8E+75$) and represents the maximum single 'mass' that can exist without generating a Big-Bang.
Universal Period	The time elapsed since the last Big-Bang or between subsequent Big-Bangs
Viscous	Solid or liquid matter in which magnetic field energy is greater than an electron's electrical charge

All other definitions can be found on the following web page:
http://calqlata.com/help_definitions.html

The Theory of Spin

A3 Symbols

The mathematical symbols used in this book are listed below:

CE	Centrifugal energy
e	Elementary charge unit
e'	Proton charge
e_m	Maximum proton charge
E_0, E_1, E_2 & E_3	Component spin energies
g	Gravitational acceleration
J	Polar moment of inertia
k	Coulomb's constant
KE	Kinetic energy
KE_e & KE_p	Kinetic energy (of electron & proton)
KE^A & KE^P	Kinetic energy (@ apogee & perigee)
δKE	(KEP - KEA)
m	Mass
m_1 & m_2	Mass of force-centre and satellite
m_e	Mass of electron
m_p	Mass of proton
N_n	Number of neutrons
N_p	Number of protons
PE	Potential energy
PE^A & PE^P	Potential energy (@ apogee & perigee)
q_1 & q_2	Electrical charge of force-centre and satellite
r	Body radius
r_e	Radius of an electron
r_p	Radius of a proton
R	Orbital radius
R_e	Orbital radius of an electron
R_p	Orbital radius of a proton
R_n	Neutronic radius
SE	Spin energy
SE_e	Spin energy in an electron
SE_p	Spin energy in a proton
t_o	Orbital period
t_n	Neutronic (orbital) period
t_s	Spin period
T	Temperature
T_n	Neutronic temperature
v	velocity
X	Velocity heat transfer constant

The Theory of Spin

Mathematical symbols continued

α	Angle of satellite tilt
β	Angle between true and magnetic north
Δ	Radial modifier
θ	Planetary tilt
ρ	Density
ψ	Neutronic ratio
ω	Angular velocity
$\omega_0, \omega_1, \omega_2$ & ω_3	Angular component velocities
$\lvert ?? \rvert$	Modulus (positive) value

Common subscripts are listed below:

c	Core
m	Mantle

A4 Orbital Motion (formulas)

Whilst this publication is dedicated to spin theory, Newton's mathematical laws of orbital motion are included here for information because you need Newton's orbital energies to calculate spin.

Sym	Description	units
t	Orbital period	s
R^P	Radius at the orbital perigee	m
θ	Any angle in orbit from apogee	°
R^A	Radius at the orbital apogee	m
m_2	Satellite mass	kg

Table A4.1: *Input Data*

Sym	Formula	Description	units
R	$p / [1 - e.\cos(\theta)]$	Orbital radius at θ	m
a	$(R^P + R^A) / 2$	half the major axis of the ellipse	m
e	$-R^P + \sqrt{[\,R^{P2} - 4.a.(R^P-a)\,]} / 2.a$	eccentricity of the ellipse	
b	$\sqrt{[\,a^2.(1-e^2)\,]}$	half the minor axis of the ellipse	m
p	$a.(1-e^2)$	half-parameter (of orbital path)	m
f	R^P	focus distance (orbital perigee)	m
x'	$a - f$	distance from focus to ellipse centre	m
A	$\pi.a.b$	orbital swept area	m²
L	$\pi.\sqrt{[\,2.(a^2+b^2) - (a-b)^2/2.2\,]}$	orbital path length	m
K	t^2/A^3	orbital constant of proportionality	s²/m³

Table A4.2: Orbital Shape

Sym	Formula	Description	units
m_1	$\varphi.(2\pi)^2 / G.K$	Force-centre mass	kg
m_2	input	Satellite mass	kg

Table A4.3: Masses

Sym	Formula	Description	units
v^P	$2.A / t.R^P$	satellite velocity at orbital perigee	m/s
v, v_c	$2.A / t.R$	satellite velocity at θ	m/s
v^A	$2.A / t.R^A$	satellite velocity at orbital apogee	m/s
g^P	$-v^P.v^A / R.(1+e)$	gravitational acceleration at perigee	m/s²
g	$-v.v^A / R.(1+e)$	gravitational acceleration at θ	m/s²
g^A	$-v^P.v^A / R.(1+e)$	gravitational acceleration at apogee	m/s²
F	$-g.m_2$	gravitational force from force-centre	N
F_c	refer to Chapter 3.2.7	centrifugal force in satellite	N
PE	F/R	potential energy between bodies	J
KE	$\tfrac{1}{2}.m_2.v^2$	kinetic energy in satellite	J
E	PE + KE	total energy	J
h	$R.v$	constant of motion	m²/s

Table A4.4: Orbital Performance

A description of the above formulas can be found in my earlier publication:
Philosophiæ Naturalis Principia Mathematica Rev. IV; 978-1-07215-605-5

The Theory of Spin

A5 Useful Formulas

Equidistant arc-length between 'n' points on the surface of a sphere:
d = π.A / C.n
where C is the circumference of the sphere
Linear distance across arc-length 'd' (above):
ℓ = 2.R.Sin(½.d/R)
but if you know 'ℓ' and need to find 'n':
n = π / Asin(½.ℓ/R)
and if ℓ=R:
n = π / Asin(½) = **6**

Lorentz's Equation (magnetic force or field strength):
F = q.v.B
Which becomes:
F = q.g.R.B
for the laws of orbital motion
Where:
q is the total electrical charge = $q_1.q_2$ / $m_e.(q_1+q_2)$
v = relative velocity (electrical circuits)
g = gravitational attraction between m_1 & m_2
R = radial separation between m_1 & m_2
B = $\mu_o.e/R_n$ = $R_n.m_e/e^2$. e/R_n = m_e/e = 1/RC kg/C
RC is the relative atomic charge capacity of an electron {C/kg}
B = 1/RC = 5.685634367312E-12 kg/C

Inter-atomic force factor (F_T):
T_k = T_n / $\xi_m.Y^2$
F_T = T_J/T_k

T_J = measured temperature of atom (shell-1 temperature